JESSE BARKSDALE

WORMWOOD

PLANET X is this "Wormwood ?"

I

AUTHOR'S NOTES

I have heard many statements about the end of the world, and the coming of the Lord on a particular year. Some even say the month and day the Lord will come and take all of the righteous back to heaven with him. I'm sorry to say no one knows when the Lord will come, or the end of the world.

I decided to write this book because I have been hearing a lot about Planet X/Nibiru that supposed to have come and destroyed the earth. There are believers out there that think this is true and all civilization will be wiped out.

The Planet has been in space for more years then anyone can count, and it will continue to be there. There is nothing man can do about it, but the good news, it will not destroy the earth. Only God will destroy the earth and no one knows how it will be done. There are scriptures in the Bible that will give you an idea as to what will happen. It will not happen like man think it will happen.

III

DEDICATION

This book is dedicated to the one who doesn't believe everything they hear, but do their own research. Look for truth and be honest during your research, and if the information is wrong, let it be wrong, and if it is right, let it be right.

CONTENTS

Author's Notes..P-2

Dedication..P-4

ACKNOWLEDGEMENTS..P-8

Chapter One Nibiru/Planet X...P-10

Chapter Two Anunnaki...P-13

Chapter Three Wormwood..P-17

VII

ACKNOWLEDGEMENTS

I appreciate my wife who is my best friend , and I thank God for her knowledge of God's word, and her contrbution to the writing of this book,

I would like to give credit to all of the information that were taken from articles written by the one I quoted in this book.

Also, I would like to thank The National Space Aeronautics and Space Administration for the photos in this book.

Chapter One

NIBIRU / PLANET X

There is a rouge planet that has entered into our solar system. The National Aeronautics and Space Administration (NASA) have known about the planet for years. The planet is called "Nibiru", the name given to a mystery distant planet in out solar system that has a 3,600 year orbit around our sun. The planet also has been called "Wormwood", or "Planet X", and NASA says the planet has entered into our solar system. This statement was made last year in June, 2015 and that is when the weather on earth began to change for the worst.

Many people have asked questions about the weather? Strange weather has begun to effect the earth; such as earthquakes, storms, tsunami, cold one day, and hot the next. The ice is melting at the North Pole, and scientists are getting worried that this planet will cause problems for us on earth.

NIBIRU / PlANET X

"This strange planet is believed to be twenty times bigger than Jupiter, with a burning moon which acts like Nibiru's personal sun. Since Nibiru goes much, much further away from the sun; this theory actually does make sense, and stands out" according to sicentific facts.

There are many who believed that the Government has been tracking Nibiru, or 1Planet X for many decades and in order to avoid chaos and unrest they have a Bug-Out Plan. They have been doing everything they can to cover-up and put out information to discredit, creditable

NIBIRU HEADED TO EARTH

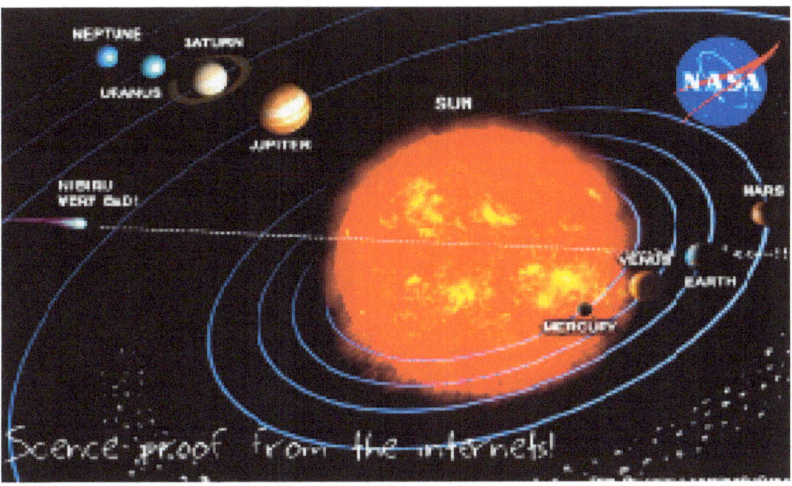

accounts of tracking and sightings.

"One of the key pieces that has been discussed are information regarding JPL's deep space Infrared Electronic Telescope photographs of the Planet X in 1981.

1. Planet X is a term used to describe hypothesized planets, orbiting the Sun beyond Neptune (the 8th planet)

Jesse Barksdale

This information has been released to the media.

There is an Ancient Prophecy that some believe about the 10th Planet. The Pope and others in the Vatican knows about this Prophecy.

There are those who believe the Prophecy about the "Anunnaki", and the return of the Planet Nibiru/Planet X to earth in the 21st. Century around the year of 2015 and 2016. (Information in the articles or in the Public Domain', and from the National Aeronautics and Space Administration (NASA). articles by Sherry Shriner)

Now, there are some that say were the Sumerians Aliens one of the first ones on earth were supposed to have left information about a 12th. Planet.

None of these facts are in the KJV of the Bible.

M O O N   A T   N I G H T

Chapter Two
ANUNNAKI

Found In Genesis 25:32

Also, knoen as Jedi, or Nephilim

This is what some believe happened in Genesis 25, but Genesis 25 is not about Aliens.

"Who were the Anunnaki

"They were the Watchers. Watchmen assigned to earth to watch over Yahweh's Creation in the Garden of Eden and Earth. They were created by God as perfect Angels. The Watchers rebelled against God and their assignment and a mutiny followed as they abandoned their mission to oversee humans and began to defile the women of the earth by having offspring with them.

This led to a giant defect in the DNA of the offspring and an eventual almost complete contamination of the human DNA. by the time Yahweh destroyed the world with a flood because to this contamination only one family on earth was left with pure human DNA. But even after the flood, the Watchers kept revolting and even more were punished and cast out of heaven losing their first estate and habitation, as they continued to defile women and human DNA **(Genesis 6:4)** This hybridization and corruption of the human DNA is still very much a part of our world although the giant defect has been corrected and most hybridization goes undetected. Why is the church silent on UFOs, Aliens, abductions, implantations, and forced breedings?

Anunnaki Are Not Our Creators!

This is another belief in the strange Prophecies that some believe in the scientific community. "This is another belief that some have. They said that the Annunaki themselves were created beings by God in heaven and were assigned to watch over the earth. When they rebelled, they were cast out of heaven, their first estate, although they do still reign in the first and second heavens and inhabit other planets and star systems visiting Earth in UFOs. Many of them have underground bases here in the earth. In the ensuing years the Theory of Evolution will be discredited from the very founders themselves and their pawns. They will then promote through Government Disinfo Scientists that mankind was created in a test tube by these Anunnaki and that these Anunnaki are our creators. This is part of the grand delusion and lie at the end of days.

"These Anunnaki are also known as Nephilim, and several other names. Instead of preaching the truth, our churches changed the truth to lies and preached the "Sons of Seth" facade changing the truth of Scriptures.

"These angels were the Sons of God who rebelled

against Him. Our churches also took out the Book of Enoch out of the Scriptures to hide the worldwide hybridization and the truth as to who these aliens are and what they are doing. To read the Book of Enoch and learn the truth, (Enoch book 1) It is written that the Mayen predicted the existence of Nibiru, or according to them, a certain dark energy in shape of a planet which would be coming near earth in the distant future, and evey time this planet came around, entire civilization from planet earth were wiped out.

The Prophecy says that the Anunnaki supposedly be the citizens of Nibiru that came to earth around 25,000 years ago, gave a lot of knowledge and details to the development of humanoids; whom at the time didn't have the brain capacity or thinking power to comprehend what the Anunnaki were saying. Now, the question must asked. Who created man? There are some that think that they gave instruction to God when he created man." (Sherry Shriner) There are no biblical facts to support this Prophecy.

"James McCaney, an expert on Planet Nibiru and Mayan history explained, around 10,000 years ago major devastation occurred which destroyed many civilization on our planet. He also explained who ruined cities in South Anerica that vanished not because of war or plague, but major physical destruction on earth.

He also went on to say that before Higiru passed us by 10,000 years ago, the North Pole was somewhere in the state of Wisconsin, while the South Pole was somewhere in the Pacific Ocean. If this is right, and this event did occur because of Planer X or Planet Nibiru then we shouldn't worry about it for the next 740,000 years or so, right? Wrong, remember even if Nibiru crossed its path from between Jupiter and Mars, it is now surging upwards to make its longest route around the sun.

This is due to the facts that its elliptical orbit goes in a round about which is very close to the sun on one end, while 80% orbits is far away from the sun.

This is why earthquakes are happening in Japan, Chile and other places that could be due to the fact that magnetic pull from Nibiru is increasing as it nears our planet. The pull from Nibiru will increase gravitational forces of each planet in a rubber band effect." (Most of this information is not provided in Wikipedia and some information are from NASA in the Public Domain)

NASA BLUE MARBLE

Chapter Three
WORMWOOD

D on't believe the hype! "Wormwood" has nothing to do with Planet X / Nibiru that the National Aeronautics and Space Administration and the science community keep talking about it. They sometime call the rogue planet "Wormwood" because of a statement in the Bible that Christ stated in **(KJV Revelation 8 Chapter and verse 10-11)**

Vs 10 "And the third angel sounded, and there fell a great star from heaven burning as it were a lamp, and it fell upon the third part of the rivers, and upon the fountains of waters:

Vs 11 And the name of the star is called Wormwood: and the third part of the waters became wormwood; and many men died of the waters, because they were made bitter."

This is strange for scientists to call Planet X wormwood, and many scienticts don't believe in the Bible. Why then do they say that this star is maybe wormwood?

The name wormwood were one of the first names given, but later they called the star by others such as Nibiru, Planet X, Teye Marrs, and confused the planet with

Jesse Barksdale

Hercolubus. and Tyche.

A few years ago Comet Ison came close to earth, and some believed that was a planet that could do great damage to the earth. The comet passed the earth and when it neared the sun it was destroyed when it passed the sun. Oh, they said it must have been a dust cloud.

DON'T BELIEVE THE HYPE!

This is terrifying to many people who do not study, and do their own research with unstanding about what is happening to the world. I think you will find that we are living in the safest times on earth. We have been blessed to have things that never entered into the minds of our grandfathers.But yet there is danger everywhere because man is at an unrest and this is because of how we act toward each other. We kill things that we can't eat! We kill each other, and think it is right.We laugh, and cry! We continued to go and act like nothing is happening to the world. When I say, the world. I'm talking about man, and not the lower animals, or vegetation which man is destroying as he multiply on the face of the earth.

There are those who say there is no God, Until they get sick! They say, Lord help me! When they get well, they curse God to his face! This is sin that hsa taken over the world! The Evil One has caused harm to mankind, and we are falling into the Satanic Traps he has set, and it is causing us to sin before God! *Do not Believe the hype about the star falling! It is Planet X sometimes called "Wormwood" that will destroy the earth! Wormwood is about whats happening and what will happened because of sin in the world!*

"The book of revelation was written by Apostle John (KJV Revelation 1:1,9;21:2; 22:8;

18

See the introduction to John's gospel and 1 John) The title of the book describes the content, and purpose of John's writing.

Revelation was addressed to the churches of Asia Minor) Rev. 1:4) specifically mentioned in chapter two and three. The book was written at a time when these churches were under great persecution and difficulty. The most important such periods were during the reign of Nero in A.D. 37-68 and Domitian in A.D. 81-96.

The word "Wormwood is found in the 8 chapter as I stated in the chapter of this book on page 17, and it was a name of a falling star.

WHAT IS WORMWOOD?

What does the Bible say about wormwood? If you want to find what the Bible says you must cross reference.

SWEET WORMWOOD
(artemisia_annua)

Jesse Barksdale

DRIED WORMWOOD

'The name wormwood is a name for a group of plants,(herbs) and some look like scrubs, small trees, and potted plants. They can be used for a variety of things such as medicine and can be mixed to make a Cocktail called Absinthe. Wormwood Grande is one of the three main herbs used in production of Absinthe."

A BOTTLE OF ABSINTH
(Price $145.00 a bottle)

21

THE THREE MAIN HERBS PLANTS

1.Grande wormwood, one of three main herbs used in production of absinthe

2. Green anise, one of three main herbs used in production of absinthe

3. **Sweet fennel, one of three main herbs used in production of absinthe**

Absinth called "the holy trinity", and I think this turm reference to the Godhead. This is a term man made up be-cause of the word wormwood.

The word wormwood can be found in the Old Testament. In the book of **(KJV Jeremiah 9:15) God said these words. "Therefore thus saith the Lord of Hosts, the God of Israel; Behold, I will feed them, even this people, with wormwood and give them water of gall to drink"** The meaning of worm-wood, and gall is a curse on the people because of sin. Wormwood is a poison in its natural state (an unused root.

Also, in **(KJVJeremiah) 23:15** the word wormwood is used agan concerning the prophets of Jerusalem.

Man has done evil in the sight of the Lord, and because of his ways he must suffer (bitter water) which has refer-ence to trouble, and harsh treatment by his nation. Nation will wedge war on nations because of sin. People will work, play and do all kinds of evil things to each other.

"A bitter plant; bitterness"

"For the lips of a strange woman drop as a honeycomb, and her mouth is smoother than oil: but her end is bitter as **WORMWOOD,** sharp as a two edged sword. Her feet go down to death; her steps take hold on hell." (Proverbs 5:3-5)

"Heb. la'anah, the Artemisia absinthium of botanists. It is noted for its intense bitterness (Deuteronomy 29:18 ; Proverbs 5:4 ; Jeremiah 9:15 ; Amos 5:7). It is a type of bitterness, affliction, remorse, punitive suffering. In Amos 6:12 this Hebrew word is rendered "hemlock" (RSV, "wormwood"). In the symbolical language of the Apocalypse (Revelation 8:10 Revelation 8:11) a star is represented as falling on the waters of the earth, causing the third part of the water to turn wormwood.

The name by which the Greeks designated it, absinthion, means "undrinkable." The absinthe of France is distilled from a species of this plant. The "southern wood" or "old man," cultivated in cottage gardens on account of its fragrance, is another species of it."

Cycles of destruction
28th January 2016

The Ouarkziz Impact Crater in Algeria, which was formed by a meteor impact less than 70 million years ago, when dinosaurs still walked the Earth. Image credit: NASA/Public domain.

1. Impact and extinction cycles

2. Modern results

3. Causes

3.1 Planet X

3.2 The Sun's orbit through the Galaxy

1. Impact and extinction cycles

Over 180 impact craters have been identified on Earth, and most of these were discovered in the first half of the 20th century.

Impacts have been associated with mass extinctions events since the 1980s with strong evidence coming from the Chicxulub Crater, which was linked to the extinction of the dinosaurs about 65 million years ago.

Mass extinction events were soon shown to be periodic, and this led astronomers to search for periodicities in impact events.

It's difficult to determine if either of these events are periodic because both data sets are incomplete, and contained both periodic and non-periodic data. Timescales are often only approximate, and results are based by the fact that more recent events are easier to identify.

This has led to inconsistent results with some studies finding no evidence of a periodicity in mass extinctions or impact events, and some reporting periods ranging from 25-30 million years for mass extinction events, and 25-35 million years for impact events in the last 260 million years. This is around the time of the Permian mass extinction, when about 96% of species on Earth were wiped out."

Artist's impression of an impact event on Earth. Image credit: NASA/Don Davis/Public domain.

2. Modern results 1

"In a recent paper published in Monthly Notices of the Royal Astronomical Society, Professor of Biology Michael R. Rampino and climate scientist Ken Caldeira provided further evidence for a 26 million year impact event cycle using data from the Earth Impact Database 2015.

Rampino and Caldeira considered all impact craters from the last 260 million years with ages that have error bars of less than 10 million years. They also discounted craters younger than 5 million years in order to prevent the relatively large number of recent craters from skewing the results. This left 37 impact craters.

Rampino and Caldeira then used a circular method

of spectral analysis, which 'wraps' the time series in a circle with a circumference equal to the trial period. If a correct period is found, then, in a perfect dataset, all of the data will be in the same place on the circle. If there is no period, then they will be distributed randomly.

Their results show a period of 25.8±0.6 million years. The latest impact event occurred 11.8±1 million years ago, and so if this cycle is correct, there should be another impact event in 14±1.6 million years.

Rampino and Caldeira use the same method to determine a period of 27.0±0.7 million years for mass extinction events in the past 260 years. The latest mass extinction event occurred 16.0±1.3 million years ago, which means we are due another in 11±2 million years.

There are 11 main impact events, and 10 mass extinction events, where 6 of these appear to correlate.

3. Causes 1

It is still not known why impact events appear to be cyclical. The most likely suggestions are that they are caused by a massive undetected object in the Solar System, or by the Sun's movement through the Galaxy.

3.1 Planet X 1

In the 1980s, it was suggested that impact events occurred in cycles because of a massive undetected object in the Solar System that periodically passes close to the Oort Cloud, the sphere of comets that orbit the Sun. This may cause comet showers throughout the Solar system.

This object could be another star, referred to as 'Nemesis' or the 'Death Star', which would make the Sun part of a binary system.

This has since been disapproved, however it's still pos

sible that there's a planet sized object in the 2 Oort Cloud or possible that there's a planet-sized object in the 2 Oort Cloud or Kuiper Belt that periodically causes impact events.

The Kuiper Belt is a belt of asteroids that orbits the Sun from beyond Neptune, and so this theory predicts that most impact events are caused by asteroids and not comets. It has been suggested that most large craters are caused by comets, however results have been mixed.

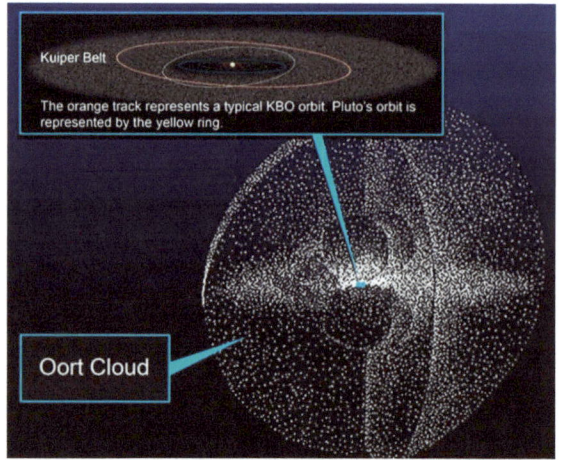

Image credit: NASA/Public domain.

A potential massive planet in the Kuiper Belt is referred to as Planet X. It was first suggested that Planet X would have to be about 5 times the mass of the Earth, which is about 1/3 the mass of Uranus, the least massive of the outer planets. However it is now thought that it could be around

2.   The outer Oort cloud is only loosely bound to the Solar System, and thus is easily affected by the gravitational pull both of passing stars and of the Milky Way itself. These forces occasionally dislodge comets from their orbits within the cloud and send them toward the inner Solar System.

the same mass as the Earth if it happens to be in the right place.

Astronomer Mike Brown and planetary scientist Konstantin Batygin have recently shown that there may be a planet at the edge of the Kuiper Belt that is around 10 times the mass of the Earth, and astronomers are now looking for evidence of this.

3.2 The Sun's orbit through the Galaxy 1

Other early suggestions involved the Sun's path through the Galaxy. The Sun does not orbit the Galactic centre in a flat plane, the way that planets orbit the Sun. Instead, it's thought to move up and down, in a wave pattern.

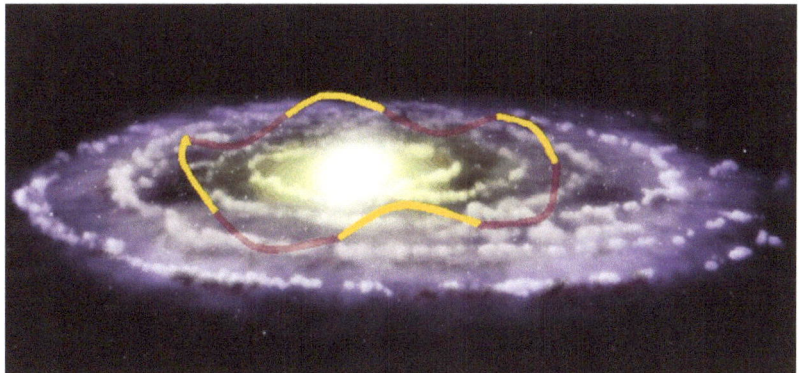

Diagram showing the orbit of the Sun through the Milky Way (approximate and exaggerated for clarity). Image credit: modified by Helen Klus, original image by NASA/CXC/M.Weiss/Public domain.

The Sun takes about 230 million years to orbit the Galaxy, travelling at about 250 km/s (just over half a million miles per hour). It passes through the vertical gravitational centre of the Galactic disc once every 30 million years or so. The Sun moves up and down because it is pulled by gravity.

It is pulled towards the centre of gravity, overshoots slightly, and is then pulled back up or down. It moves about 200 light years from the centre of the 1000 light year wide disc at each maximum or minimum. The Sun passes through the centre about 8 times with every Galactic orbit.

It was first suggested that the Oort Cloud was affected as the Sun moves through the gravitational centre of the Galactic disc. In 1996, Rampino referred to this idea as the Shiva Hypothesis after the Hindu God of Destruction.

One problem with this idea, however, is that scientists think we last moved through the centre of the Galactic disc about 1 million years ago, yet the last impact event was about 11 million years ago.

This idea has since been extended by physicists Lisa Randall and Matthew Reece who showed that the Oort Cloud may be affected in a similar way if it passes through a thin disc of dark matter.

If the Earth passes through a dark matter disc, then it may capture some of the dark matter. Dark matter could form a ball at the centre of the Earth's core, causing the core to increase in temperature.

This may lead to an increase in volcanic activity and earthquakes, and could even cause the Earth's magnetic poles to reverse, so that the North Magnetic Pole becomes the South Magnetic Pole, and vice versa.

There's currently no strong evidence for a dark matter disc around the Milky Way, however Randall and Reece's predictions can be tested by the ESA's Gaia satellite. Gaia is currently mapping the gravitational field of the Galaxy, and the first set of results are due to be released this summer.

If a dark matter disc is detected, then it would mean that the geological and biological evolution that has taken place on Earth is directly linked to the distribution of matter in the Galaxy."

Maybe Planet X is out there somewhere, but I would not worry about it because the world will not end until God say it will end. The only one that knows when the world will end is God. When you hear the world is coming to an end, do not believe it because it is not true.

No planet will hit the earth and destroy it!

We will have changes in the weather, and man will fight and kill each other, Man will start wars and many will be killed in warfare.

All I can say is do not believe the hype!

www.ingramcontent.com/pod-product-compliance
Lightning Source LLC
Chambersburg PA
CBHW041613180526
45159CB00002BC/836